BLADESMITHING 101 FAQ

Answers to Your Burning Knife Making Questions about Forging, Stock Removal, Tools, and Heat Treatment

WES SANDER

TABLE OF CONTENTS

INTRODUCTION

B ladesmithing is a fun and complex hobby that many people enjoy. The thrill of making your own knives is something that most individuals love to experience. Not to mention, it is a great learning experience.

However, when I work with new knifemakers, I notice that they often have questions that come up quite frequently. That is why I have compiled 101 important questions that you may have about bladesmithing and answered them for you.

We will also discuss the tools you need, how to forge, and even how to use recycled steel to create usable blades. Furthermore, we will talk about the properties of various metals, especially steel. Understanding the tools of the trade is important when bladesmithing. Otherwise, the process can be frustrating.

I will highlight a few examples in the book regarding different smithing techniques. They will allow you to know what is best for every situation that you may encounter in the workshop.

FREE BONUSES FOR THE READERS

F irst of all, I want to congratulate you on taking the right steps to learn and improve your bladesmithing skills, by buying this book.

Few people take action on improving their craft, and you are one of them.

This book has exhaustive knowledge on bladesmithing and will help you make your first knife.

However, to get the most out of this book, I have 3 resources for you that will REALLY kickstart your knife making process and improve the quality of your knives.

Since you are now a reader of my books, I want to extend a hand, and improve our author-reader relationship, by offering you all 3 of these bonuses for FREE.

All you have to do is visit **https://www.elitebladesmithing-masterclass.com/free-bonus** and enter the e-mail where you want to receive these resources.

These bonuses will help you:

1. Make more money when selling your knives to customers
2. Save time while knife making

Here's what you receive for FREE:

1. Bladesmith's Guide to Selling Knives
2. Hunting Knife Template
3. Stock Removal Cheat Sheet

Here is a brief description of what you will receive in your inbox:

1. Bladesmith's Guide to Selling Knives

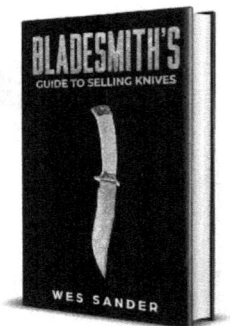

Do you want to sell your knives to support your hobby, but don't know where to start?

Are you afraid to charge more for your knives?

Do you constantly get low-balled on the price of your knives?

'Bladesmith's Guide to Selling Knives' contains simple but fundamental secrets to selling your knives for profit.

Both audio and PDF versions are included.

Inside this book you will discover:

- How to **make more money** when selling knives and swords to customers (Higher prices)

- The **hidden-in-plain-sight** location that is perfect for selling knives (Gun shows)

- Your **biggest 'asset'** that you can leverage to charge higher prices for your knives, and **make an extra $50 or more** off of selling the same knife.

- 4 critical mistakes you could be making, that are **holding you back from selling your knife for what it's truly worth**

- The ideal number of knives you should bring to a knife show

- 5 online platforms where you can sell your knives

- 9 key details you need to mention when selling your knives online, that will increase the customers you get

2. Hunting Knife Template for Stock Removal

HUNTING KNIFE TEMPLATE

Tired of drawing plans when making a knife?

Not good at CAD or any sort of design software?

Make planning and drawing layouts a 5-second affair, by downloading this classic bowie knife design that you can print and grind on your preferred size of stock steel.

Here's what you get:

- Classic bowie knife design **you can print and paste** on stock steel and start grinding
- Remove the hassle of planning and drawing the knife layout during knife making
- Detailed plans included, to ensure straight and clean grind lines

3. Stock Removal Cheat Sheet

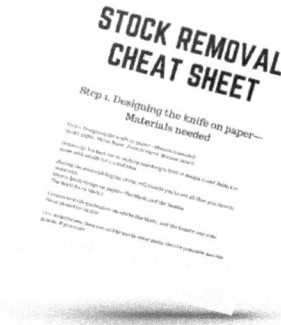

Do you need to quickly lookup the correct knife making steps, while working on a knife in your workshop?

Here's what you get:

- Make your knife through stock removal in just **14 steps**
- Full stock removal process, done with 1084 steel
- **Quick reference guide** you can print and place in your workshop

As mentioned above, to get access to this content, go to *https://www.elitebladesmithingmasterclass.com/free-bonus* and enter the e-mail where you want to receive these 3 resources.

DISCLAIMER: By signing up for the free content you are also agreeing to be added to my bladesmithing e-mail list, to which I send helpful bladesmithing tips and promotional offers.

I would suggest you download these resources before you proceed further, as they are a great supplement for this book, and have the potential to bring an improvement in your results.

CHAPTER 1: TOOLS OF THE TRADE

Tools are your first order of business. We will highlight some common questions related to this below.

What Is a Good Arc Welder for Beginners?

The ESAB Thermal Arc Inverter Welder is an excellent option for beginners and more experienced bladesmiths.

It is great for TIG welding and weighs 10 pounds, making it one of the lightest arc welders on the market. The tool comes with a one-year warranty and delivers about 95 amps of power, so it is suitable for basic projects.

The Miller Electric 120/240 AC Welder is another good choice that new knifemakers can use. It is a simple and beginner-friendly welder that you can start using right away. It is suitable for welding steel around the thickness of ⅜", so it is perfect for the small- or mid-size steel.

The Miller Electric 120/240 AC Welder is among the most versatile and dependable tools out there. The only downside is the fact that it is one of the pricier models on the market. But if you think you can use it, you should pick one up.

How Much Should I Pay for an Anvil?

The amount depends on what you are getting. On average, approximately $2 to $5 per pound is a reasonable price for an anvil. So, for a 100-pound anvil, expect to pay anywhere from $200 to $500.

Some anvils can range from $7 to $10 a pound, but they rack up a pretty penny after a certain weight. For beginners, therefore, I do not suggest buying them. Most basic smiths start with 150-pound anvils, so you may need to page within the $400-$900 range for one.

Alternatively, you can get secondhand anvils. However, you should look for signs of wear and tear, especially on the edges. You should not put money down until you know for sure that you want to stick with it.

You can even make an anvil for yourself, using a piece of railroad track.

How Much Should I Pay for Calipers?

On average, calipers range from $30 to $70 for a small set. A lot of beginner smithing sets will sell them with other tools that you need for the job. The price can then be around $200 or $300 because of that.

Another thing you can do is make your own calipers.

Meanwhile, if you buy used calipers, they come at a pretty affordable price. You just have to make sure that they have the proper grip. Oh, and try not to buy brake calipers because they cost way more and are not for blacksmithing.

What Oil Should I Use to Maintain My Leather Sheath?

Dovo leather balm and Ballistol are excellent oils to use for maintaining leather sheaths. You rub on the former and then wait about 15 minutes for it to settle. The sheath will be nice and lubricated with either of these options.

I do not recommend using shoe polish because it rubs off on clothing, staining it black. The only exception is if there is a clear coat on top to seal it or at least sno-seal to keep the dye in place. Then, you can apply it and use a hairdryer to seal the shoe polish.

Mink oil is another good option if you want a more natural oil that does not have a smell. It waterproofs your gear as well, which is perfect for a survival or fishing knife sheath that you will be using on the water or in wet places. It also does not take long to dry after the application.

When choosing an oil, you want to make sure that you can use it to keep everything in fantastic shape. Some of the choices mentioned above can be used on the blades, but they're better for lubricating, strengthening, and keeping the sheaths shiny.

How Do I Make My Own Scales?

The simplest way to make scales is by following the Micarta process. You will need resin and fabric to do it.

First, cut the fabric and line it with wax paper. Mix the resin while wearing a respirator to avoid inhaling harmful chemicals. Once the resin is done, apply it on the fabric layer by layer and saturate it. Then, fill the top and use a clamp to keep everything in place.

Let the resin cure before unveiling your new scales.

What Are the Different Parts of a Knife?

The different parts of the knife include the following:

- Point - used for piercing

- Tip - used with the knife for cutting and forcing
- Spine - the opposite of the tip and is usually duller
- Edge - the cutting element next to the tip, across from the spine
- Heel - the back of the blade, opposite the point
- Bolster - the part that connects the knife to the handle
- Tang - the section next to the bolster that extends to the knife's end
- Scales - the part that makes up the wooded area of the blade
- Rivets - the metal pieces attaching the scale and the tang
- Butt - the end of the knife

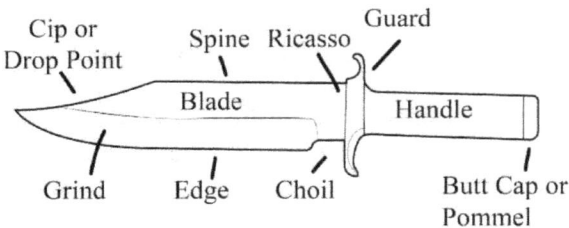

How Do I Start Making A Knife With Only Hand Tools?

It is possible to make your knife with just a few tools.

You can get an angle grinder, sandpaper, files, clamps, drills, a Dremel, and even a grill to heat the knife. You can make a pattern and then grind out the blade in your steel. Just be careful and always wear goggles. Steel slivers in your eye can never be fun.

You can grind out every edge as well and drill the holes for the rivets to keep the tang and the handle together.

Later on, you need to heat treat and shape your scales.

Sharpen your knife with the flat shape of your bevel but do not make it too sharp.

What Is a Good Groove Tool for Leatherwork?

A leather groover stitcher is a wonderful tool, especially if you want to sew your sheaths without using score lines. The Versa Groover by Bob Douglas is a good option for leather grooving, as well as the Pro Stitching Groovery by Tandy Leather.

The latter, to be specific, can be used to measure out and figure out your leather grooving options. It can also help you work out an exact pattern to create beautiful results. However, I do not suggest sewing leather by hand because it is time-consuming and incredibly hard.

Are Engine Parts Good for Making Blades?

That depends. If you have free parts lying around, such as some axles, leaf springs, steering linkages, and torsion bars, then it is worth it. However, if you are raiding a junkyard, it will cost you money that you may have used to buy steel online for the same or cheaper price.

Unknown steel heat treatments are almost always trial and error as well, so you should take your time with these. If you are going to use car parts for this, then I suggest knowing what alloys are in it. Still, it may not be the best idea for beginners.

How Do I Set Up a File Jig?

You put the file underneath the rod but over the knife. To change the file that you are using for grinding, as well as the angle, you can choose which hole it is for.

You can adjust the tapering at this point, but you can set this up with some file pieces, L-shaped brackets, and utility steel from the hardware store. Then, put the knife onto the wooden block on the vise itself. You can secure it over a tabletop, too. Just understand that how you use it will determine the placement of the blades.

What Are the Most Hardwearing Handle Materials?

If you want a long-lasting and robust handle, then I highly recommend getting titanium. It is the toughest and hardest material out there. Only, it happens to be the most expensive as well, so you want to watch out for that.

Synthetics make excellent handles, too. They are usually waterproof, so I suggest using one if you are making a fishing knife. The issue that comes with man-made materials, though, is that they're not as sturdy and resistant to wear and tear like the others.

Why Is Titanium Usually Used for Scales on Framelock Knives Instead of Mild or Tool Steel?

First and foremost, titanium is incredibly strong and durable for its weight. It does not rust either, which is why many people find it marketable. As expected, it is the most expensive material that you can use to make scales.

Yes, you can make titanium knives, too, but it is a lot more expensive than other options. Besides, they work better as scales than anything else. Framelock knives need a stronger material like titanium, but you can use tool steel for the blade itself.

Is There Any Downside to Using Lighter (and Cheaper) Anvils?

For smaller knives, you do not need some giant 400-pound anvil. One that's about 100 or 200 pounds should suffice. So, you can go a little cheaper on this, depending on what type of project you are doing.

Most people can get the job done on a 150-pound anvil, too. To put it into perspective, master blacksmith Francis Wittaker used a 150-pound anvil all his life. It entails that you can create a lot of things on a smaller anvil.

If you are working with larger knives, though, a lighter anvil will typically take much longer to forge. Meanwhile, a heavier anvil will rebound the hammer, helping with shaping. You can get more work done on them with a bit less effort as well.

Remember: Bigger is not always better, but you sometimes need to keep in mind that the tool should always be relative to your project.

Is Beeswax a Good Compound to Prevent Blades From Rusting?

Beeswax won't protect the blade from rust. However, putting wax on cutting surfaces does help protect them. The best way to prevent rusting is by wiping it completely dry when you are done using it.

The problem you may notice with beeswax, however, is that it can trap lint and dust. Hence, it can harm the blade in that regard. Some people like to use beeswax despite that, but it all depends on your preferences.

Can I Heat Treat a Small Blade With a Torch?

That depends. If your torch gets up to the temperature you need for heat treatment, then you can heat the blade. Ideally, you want something that reaches at least 1500 degrees Fahrenheit to heat treat high carbon steel. The best items are the A/O or Presto-lite torches because they have interchangeable tips.

The biggest challenge is figuring out how to hold the blade without hurting yourself. Keep in mind that you need to heat evenly with a torch. You also need to do it in parts, so the process may take a while as well.

Try not to use torch the blade in bright lighting since you won't see if the flame is red hot. Since you are working with high heat, make sure that you have the right equipment for this. Once you are done, quench the blade in canola oil so that it gets down to the right temperature slowly.

If I Had to Buy Just One Power Tool for Knifemaking, What Would It Be?

Get yourself an angle grinder. It is probably one of the most essential knifemaking tools out there. Not only is it vital for cutting and grinding, but it is also useful for sharpening knives.

You can use an angle grinder for a spark test, which we will discuss in the book. It can even cut the steel into a more exact state, which is terrific for knifemakers. If you are on a budget and can only get one power tool, then you will want to consider buying this or a belt sander.

Another thing is that angle grinders are great for rust removal. If you have a rusty blade, it can do quick work of that, along with dull blades. It can be an all-in-one tool in your workshop.

What Tool Do You Need to Cut Serrations Into a Knife?

A Dremel will be your best friend at this point. It is best if you use a diamond blade attachment to cut the serrations into the knife. You should also have a pattern drawn where you want the cuts to be.

You should start with the smallest diamond bit, grinding away at the blade at an angle. From there, move on to bigger bits. Make sure that your grooves are centered as you do all this, though. You will have to repeat the process on the backside

What's a Good Angle to Set My Jig for Bush or Survival Knives?

The ideal level for Western blades is about 20 degrees on each side. It means that you are sharpening it at about 40 degrees. The reason is that the angle creates a bevel with a thicker spine on each side, which is what you want from survival knives.

Asian cutting knives are done at 20 degrees one side and 0 on the other side, meanwhile. If you are using an angle for a softer knife, it can be under 10 degrees. Anything over 30 degrees on either side is not beneficial for axes or huge knives.

When choosing the ideal level to sharpen and form knives, however, you need to understand its purpose. Decorative blades may need only one side beveled. Survival knives are supposed to be sharper because they need to cut and chop in the wilderness.

Angling survival knives at around 20 degrees on each side is integral. You want it to be sturdy and long-lasting but without the plunging power or the cutting depth that other knives may have.

What Are Essential Bladesmithing Tools to Get When I'm Finally Setting Up a Workshop?

The forge is probably the first thing that you will need to set up. Having an anvil is also essential, especially if you plan to forge your blades. You also have to have vises to grip the edge, as well as blacksmith tongs to grab the metal that you are working with.

Of course, you may also want to consider purchasing slitters, center punches, and twisting tools since they are used more for perfecting the skill.

In general, you can put this all together with approximately $300 to $500, considering you can find inexpensive anvils, too.

CHAPTER 2: FORGING BLADES

Forging blades comes with a series of steps. If you have questions about it, I will try to answer some of them in this chapter. This way, you can get a lot of information about the process before trying to create your blades.

How Do I Get the Perfect Plunge Line for My Knife?

First, grind the blade thoroughly. Then, set your plunges where you want them. I suggest starting with one before lining up the others based on where you put the first one.

Some people will run an 80-grit paper over the wheel to see where your plunge lines need to be. do not grind with the belt moving away from the plunge.

I also have gotten wonderful results using a 160-grit, but anything higher than that can spell trouble for the average user, since it will grind it too much, and can affect the state of the blade.

How Do I Build My Forge?

Get a stainless steel sink that has two basins at the center. Fill one side with water and then line the other with furnace cement. Cover the sink with fire bricks that do not have mortar in them.

Then, put a flood drain over the sink grain and blast the air as the coal enters the bed underneath. You will need to buy a 275-pound anvil to help with shaping.

This is the cheapest way to do it, and you can get all these tools at the hardware store.

How Can I Forge Stainless Steel?

You can forge stainless steel with specific hammers, but understand that it will take a very long time. Due to its composition, you can form this into many other shapes as well. A D-2 round rod is an excellent option for forging stainless steel.

I like forging stainless steel by taking my time and using an oil quench rather than a water quench with this because it yields better results. I also use a grinder almost every time I work with stainless steel to make the job easy.

Can I Forge Scrap Metal Into Knives? How Long Will They Last?

I do not recommend forging scrap metals unless you are working with a higher carbon tool or stainless steel. The reason is that the softer carbons do not take to forging very well. Besides, scrap metal has a lot of impurities to it.

Some metals may have chrome plating on them, which is incredibly unsafe for forging. If you are going to use scrap metal, though, the tool steel, stainless steel, and carbon steel are your best friends. Just do not expect them to last for a long time. you are using junk materials, so you may get junk blades.

Is It Better to Use Oil or Water for Quenching?

The answer depends on how quickly you want the blade to cool.

Water will quench the piece much faster since it can take the heat out of the medium more rapidly. It is also useful if you are more concerned about the hardness than the stress you are putting on it.

Oil cools things down at a much slower rate, so if you want it to lower the blade's temperature slowly, this is better. It is ideal if hardness is not as much of a concern and you are more worried about the stress that you are putting on the blade.

What Do I Do If I Do Not Want to Buy an Expensive Forge to Heat Treat My Knives?

In this situation, some people will send their blades directly out to a heat treatment company. After all, it may be more cost- and space-effective to do that if you cannot buy a forge.

However, if you have the money and space to spare, I do recommend getting a basic heat treatment forge. They are not as pricey as you expect, and there are a few that come under $200. One of the benefits of this type of forge is that you can get it up to 2300 degrees. You can use it to make more advanced knives.

If you know someone who has a forge in their home, you can also ask if you can use it. In case there is a place nearby that offers smithing classes, they should have a forge that you may be able to borrow as well.

If You Are Quenching an Item in Oil, How Long Should It Be Left There?

If you want to get technical, the TTT model will tell you what your structure should look like at each cooling stage. However, for basic knifemakers, you are okay with letting the blade quench in oil the temperature goes below 400 degrees. Some people like the TTT chart, especially if they are working with sophisticated edges.

After Quenching, Do I Remove the Scales From the Blade Before or After Tempering?

You want to remove the scales before machining and tempering your blade. Besides looking a little unsightly, it can end up flaking off and ruining your knife if you do not remove it right away.

Some people tend to use the no-scale solution to treat the blade. However, using a grinder to get rid of those unsightly scales will help you immensely. Doing it afterward can also ruin the edge.

You typically want to do the quenching and then the scale removal before you temper your blade. However, understand that these scales can form at lower temperatures, even during heat treatment. Sometimes, they may not even form at all due to the type of alloys in your metal.

Before you start removing the scales from the blade, therefore, you should always make sure to test the metal. It will give you an idea of what type of material you are working with. Thus, you can achieve the best results possible.

Should I Heat Treat the Tang of My knife?

That depends. When you heat treat the tang of a knife, take note that doing so will make it much harder to drill the holes into the tang to bring it together.

Some people do not like heat treatment because they prefer to work with a rougher tang. However, a lot of knifemakers will do this because the natives resist tempering anyways. Besides, it does not cause too much damage.

If you are going to heat treat the tang of your knife, therefore, you should definitely make sure to temper it afterward.

How Important Is Tempering After Quenching?

Honestly, it is essential to start tempering after quenching. If you do not do it, then your blade will easily break.

Many beginners will neglect the tempering step, but that will make the blades brittle and liable to breaking. After all, the process increases the hardness of the blade after you do it because you are heating the metal again just below the critical level.

From there, your blade will naturally toughen over time.

Should I Forge Anything Out of Aluminum?

You can forge an object out of aluminum if you want to. You can find this material in pistons, gears, and wheel spindles within cars, so it is readily available.

Aluminum is better to be used with a cast rather than a forge because its melting point is around 1350 degrees. This temperature is quite low in comparison with other steel types that go beyond 2300 degrees.

Many beginners like to cast aluminum since it is easier to work with than other materials. Ideally, pliers and hammers work wonderfully if you forge them out of aluminum. However, if you overwork this material, it will get brittle.

The Surface for My Anvil Is Not in Very Good Shape. Can I Take It to a Machine Shop and Have It Milled Without Damaging It? Or Should I Leave It Alone and Use as It Is?

Here's the thing: sending out an anvil somewhere to repair it can be costly. You still have to factor in the price of transporting it, which can be super expensive.

The best thing for you to do is to smooth the anvil out with a grinder. It will help keep the edges better, and you can use it for your work still.

Do not try to heat an anvil up to "fix" it because it will ruin the steel, and you do not want that. You can smooth out the edges minimally. If you do anything more or bring it to the shop, it is ultimately not worth it.

I Am Making a Knife From Old File. After I Somewhat Get It to Shape and Before Hardening It, How Sharp Should the Blade Be?

When doing this, you should make sure that the blade is sharp enough to cut everything in a downward motion easily. You can test it by lightly piercing a piece of paper.

If you notice that the blade does not cut the paper thoroughly, it is too dull. However, if you get a nice, even cut, then you have a sharp knife.

Usually, testing a blade by using it is your best solution since it will show you that it is doing its job.

Do I Drill Holes in the Tang Before or After Heat Treatment?

You need to drill holes in the tang before heat treatment if you want to make knives easily.

The purpose of heat treatment is to harden the blade. Sure, you can drill holes afterward, but it is better to do it beforehand since that's when the steel is much easier to work with.

Drilling slightly bigger holes is also advantageous for you. By doing so, they will typically be able to hold the blade in place.

You should also drill before you bevel. All of your steps will be easier to do before heat treatment.

What Should Be the Pivot Location on a Folding Knife? What Should Be Its Position?

So long as the spring is flush, and the liners are open or closed, then that is an excellent pivot location. You can also do it with closed or half stop as well if there is one.

You want to make sure that you have a smooth action when you do this. If there is a snap when you open or close the knife, then you do not need to worry about the pivot's position.

Generally, people place the pivot near the halfway point between the sheath and the blade but with a little more room to accommodate the blade's size. However, snap knives do not require this.

When Coffee Etching, Do You Use Hot or Cold Water?

Coffee etching is a process that literally etches the blade with coffee. It is a popular method. Plus, you are not working with any harmful chemicals.

The water, however, needs to be distilled water to reduce the number of impurities in it. Ideally, you want the water to be around room temperature. The reason is that hot water will turn the acids into a base, which can affect the coloring.

I do not suggest coffee etching on a chef's knife since the black nature of it will come off very quickly. Despite that, it does create a pretty color that ranges from dark black to a smoky gray, depending on how you etch with coffee.

My Blade Warped After Quenching. Can I Go Ahead and Temper It as Is and then Address the Warp Without Breaking It?

You should try to address the issue before tempering the blade. To fix a warped edge, you need to heat it up to about 1000 degrees. When you reach that temperature, you can start to straighten it slowly.

Once you notice that the blade is cooling down, do not try to continue to straighten it further. You can temper the edge to a higher temperature, but this does result in a softer edge.

Sometimes, people can do a temper before warping, but it depends on the steel and how hot it gets. If it is nice and hot, you can slightly straighten it. Still, understand that warping is not your fault. It is something that happens to blades every now and then.

You should never straighten a blade if its temperature is below 400 degrees. Otherwise, it will be incredibly hard. You can also prevent it by relieving the stress during forging so that it does not become a problem.

Is It Worth It to Use Motor Oil for Quenching?

Motor oil is often used for quenching because it is readily available and cheaper than other oils out there. However, it brings forth the problem with fumes. Unless you take the proper precautions to handle motor oil, you can inhale those noxious fumes which are not suitable for you.

You can try to wear a respirator in a well-ventilated area while quenching. The thing is, it also takes a lot longer to lower the temperature. It entails that you are subjecting the blade and yourself to the noxious fumes of motor oil more than necessary.

The best way to quench is by using food-grade or synthetic oils.

What Is a Good Oil for Quenching?

Some people get away with quenching their knives with motor oil. While it is cheaper, the downside is that it comes with potential toxins. The other popular type is food-grade oils, such as canola and peanut oil.

You may have heard of canola oil more than peanut oil, though, because you can buy it in bulk at an affordable price and often smells pretty good.

You won't have to worry about potentially harming yourself if you use canola oil. You will, however, need to preheat it up to about 130 degrees before quenching. Still, the flashpoint of these oils is much higher than others.

Do I Quench and Harden the Knife Before or After Putting the Bevel and Edge on It?

Beveling and filing are extra difficult after you quench and harden the knife. After all, heat treating tends to soften the blade and then set it after a while.

Please take the time to shape your knife before you start to treat it. Quenching and hardening are usually done at the very end since you have the knife shape then. Beveling takes place during the shaping process.

Do You Sharpen Your Knife Before or After Heat Treatment?

It is better to sharpen the knife after heat treating it. One of the reasons is that the edges can lose carbon while they are heat treated.

Aside from that, the edges might roll, too. Sometimes, they will warp on sharpened blades if you sharpen them before the heat treatment. You will then waste your time sharpening it beforehand.

Get the blade to its proper hardness before taking the right grit of paper and sharpening the knife.

Does It Make Sense to Wear a Respirator While Forging?

Yes. In truth, I do not suggest forging without wearing a respirator. No matter what kind of metals you work with, you are going to expose yourself to some form of toxins.

All metals have chemicals in them because that is how they are made. While steel is probably the least toxic of them all, chrome probably has the highest toxicity level, and aluminum is just right below it.

Also, you need to wear a respirator if you are working with metals that you do not know the composition of. More often than not, lots of people will inhale the chemicals without even realizing it.

You are always better off safe than sorry. If you are exposing yourself to a lot of toxins while forging a blade, you need to get comfortable with wearing a respirator.

How Hot Should My Forge Be When I Am Forging? At What Temperature Should the Flames Be?

The hottest temperature that a forge can reach is about 2800 degrees. Take note that it is the melting point of stainless steel. However, people usually keep it around 2300 degrees on average.

Steel melts anywhere from 2300 to 2700 degrees, depending on how much carbon and other elements are in this. The max forging temperature is around 2300 degrees since you want it hot but not too much to burn everything completely.

Usually, forging flames have an orangish-yellow color. It is a contrast from the typical red-hot flames that you may know of. Red flares are quite low in temperature compared to these as they only reach 1300 to 1400 degrees.

Of course, forging flames are still hot, so you should always wear protective clothing when working with it.

What Types of Files Are the Best When Hand-Filing Blades?

There are a lot of great files out there!

The Nicholson/black diamond files are fantastic. If you get yourself a chainsaw file, a half-round file, as well as one that cross-cuts, you will be in business!

However, understand that it is a little harder to get the right angles with your knives by hand-filing. Also, it takes a little bit more work to accomplish this job.

How Thin Should I Make the Edge Before Heat Treatment?

That depends. For a lot of O1 steel, you can usually get away with anywhere from 0.2" to 0.35" thickness. However, giving it a thicker end can be a bit too much for the piece, so you will want to keep it around 0.2" at most.

The thinness should also depend on how much you are quenching the blade and what type of quench you are using. If you have a liquid quench, for instance, you will need to do a little bit more grinding to make it thinner. An air quench, on the other hand, will require less of that.

How thin the edge should depend on the type of sandpaper you use as well. For example, the 220- or 400-grit sandpaper will give you a nice, thin edge. If you use a 60-grit paper on the side, it will be thicker.

Some people have tried thicker edges, but it blunts. It also does not stay as sharp and as good-looking as other knives.

CHAPTER 3: STOCK REMOVAL

S tock removal is another part of forging blades that most people do not know about. In this chapter, we will answer your questions about stock removal in bladesmithing and give you some tips and tricks for doing this.

What Is the Best Knife Steel for Stock Removal?

1084 is the best knife steel for beginners. It is easy to work with, has belts that are effortless to use, and can be suitable for those who haven't done stock removal before.

1084 creates blades that can be re-sharpened without much effort. It does not last forever, however, and is not corrosion-resistant. Keep that in mind when working with this kind of steel.

O1 and D2 are also good options, but they cost a little more than the 1084.

What Is Better, Stock Removal or Foraging?

Stock removal is a much more efficient process because you do not have to spend all the time hammering away at one specific piece. However, unless the pattern is already there, the blade feels almost patterned in terms of making it.

Forging, in contrast, takes a lot longer to do and requires the addition of a human element to it. That can cause either consistencies or inconsistencies in the blade, depending on what you are creating.

Despite that, stock removal needs more machinery, which can make it ultimately more expensive.

Do I Lengthen the Taper or Make the Grind First?

Always make the tang and then taper that beforehand because you want the shape to stay in place. The grind is then used to perfect the shape of the blade. If your taper is not as long as you need it to be, you will have to do it repeatedly.

It is better to do the job right the first time and make sure that your taper is at the right length. Then, you can grind out the blade shape to perfection.

Can I Use a Bandsaw With a Blade Designed for Cutting Wood and Knife Profiles Out of Wood? Or Should I Have Separate Bandsaws?

Cutting wood and knife profiles require different speeds, so you should have two separate bandsaws. The metal saw will cut the wood and the composites much more slowly. If you want something more exact, then this is it.

The wood saw will dull quite quickly if you use it to cut metal. You should only use it on a few pieces if you have to.

You can also pick up some bi-metal blades if you want to have one saw but two different types of blades. They are usually about 18-24tpi and used quite effectively in cutting the metal that you need. Hence, it is a great option, too.

What Do you do with all the Gunk in Your Grinder Tanks?

Some people have benefitted from draining this as storm-water, but the best option in that case, is to drain the water that's in it into your garden or compost, and from there, start to bin the extra stuff that's in the bucket. Some people do get a good result from burning this, but that can be up to you. You will want to check to see if it is burnable, but some people have luck with it. You can't make a Damascus out of it, because usually there is not enough for making some.

I love to take this, and from here use it in my gardens to help fertilize. That way, my plants can grow better, and I'm not wasting it. I also give it to others who have a garden and need help growing it as well.

CHAPTER 4: STEEL-RELATED QUESTIONS

S teel is one of the main items you will be using when forging blades. But did you know there are properties of steel that go along with blacksmithing that you may not know about?

In this chapter, let me answer some questions that you may have about steel.

What Is the Best Steel for Kitchen Knives?

The best steel for the job is Cromova. It has .8% carbon and can resist wear and tear. This global steel is perfect for keeping your blades incredibly sharp.

Another fantastic option if you want something a little different is Cronodor 30. It is stainless steel that has nitrogen mixed into it to resist rust and wear. The hardness level is also great for this one and does the cutting job pretty well.

Carbon Steel HRC 60 is a non-stainless-steel option, meanwhile. It is pretty good if you want a decorative knife. However, it may not be as hard as the others. It is about Rockwell C hard but also has a decent amount of strength and .8% carbon in it.

How Do You Find Out What Type of Steel Your Knife Is Made of?

The answer to this question is obvious. Only, if you buy steel from a reputable retailer, you can email them or call and ask them. Some sites have a contact form that you can use to ask the question as well.

If the steel was bought from a supplier or metal market, they should be able to tell you right away what you have purchased.

Before I begin personally, I always check on the steel I use with suppliers. This way, I know what I am working with. Failure to do so can lead you to make odd choices when working with steel.

What Is a Suitable 440c Stainless Steel Alternative?

Well, 440c is an alternative to AISI 52100, so this is your best bet if you want something like 440c but with a little less resistance.

The Mini Swagger from Gerber is not as formidable as the 440c but is more of a 440A type of stainless steel. It still does the job pretty well, even though it is a Chinese metal.

9cr18mov is another good option. But then again, it is not as strong as the 440c even if they are similar in composition.

Is Modern Damascus Stronger Than Wootz Steel Damascus?

That is a good question because Damascus Wootz steel is the original name of this type. Damascus is more exotic, considering it uses two to five different kinds of alloy.

In the past, people were limited to how much steel they could use. However, in the modern-day Damascus steel-making process, they follow more detailed steps to make this.

While this is a personal preference, a Damascus blade is much stronger than what was used back then due to the carbon added in the materials.

What Is the Best Steel for Forging Knives?

1095 steel is a right choice if you are looking to forge basic blades that allow you to create decorative pieces. It has decent functionality and impressive lifespan because of the presence of carbon that improves its hardness.

Tool Steel is excellent because it is made with durability in mind. The A2 is the strongest of the tool steel out there. It does not get as hard as the M2 or D2, but it is rough and can withstand excessive use.

The D2 is usually added to the A2, however, because it will end up corroding if you are not careful.

What Is the Best Steel for Beginners?

Out of all your options, 1075 is the best steel for beginners. The reason is that it mostly has iron and carbon in it. This steel is also very easy to heat with blowtorches and other simple heating elements.

1075 does not need a long soaking time like different types of steel. 1084 is good as well if you want something a little more durable. However, you cannot make larger, lasting objects with the latter.

What Is the Best Steel Combination for Making Damascus?

There are a few things that go into making Damascus steel. In general, though, the 1080 and the 1084 sheets of steel are your best bet. You can combine that with 15N20 and improve the strength of the blade.

Some people also like to add D2 to their alloys because it increases the steel's hardness. However, keep in mind that this adds chromium to the mixture. This element can affect the chemical composition of the blade.

The 15N20 is one of the necessary parts of making a Damascus since it adds nickel to the piece. You want at least 6 points minimally for the carbon in it to forge a durable blade. Some also have had good results with 1085, so you should see what works best for you.

Why Is Heat Treatment for Stainless Steel More Complicated Than Carbon Steel?

Stainless steel is more malleable than your average carbon steel. One possible reason is that there is nickel in the latter. Stainless steel also comes with more elements than carbon steel.

The carbon steel may have less carbon than the stainless steel as well because there may be low-alloy versions of this. It typically has 0.30% per weight, which means that it is much easier to heat treat than stainless steel.

The heat treatment numbers might be all over the place, so understand the issue that may appear when trying to treat either steel.

What Is a Good Stainless Steel With the Least Complicated Heat Treatment Process?

The least complicated stainless steel to work with has to be N690 or CPM 154.

The reason is that the N690 typically has more carbon than the others, along with vanadium. It makes up a considerable portion of hard knives. It also has the right amount of carbon and a bit of cobalt which helps to make a useful blade.

CPM 154 is the kind of stainless steel that knifemakers look for. It has an excellent edge holding is corrosion-resistant. It contains less amount of chromium than the other steels manufactured in the same way. With a hardness level of 65, though, it is one of the hardest stainless steels out there.

Not to mention, CPM 154 has a high carbon level, which is why a lot of knifemakers encourage others to use this type of steel. The treatment process for it is much more straightforward than other stainless steels of the same name, too.

How Many Steel Billets Do You Need for a Damascus?

It all depends. You can start stacking steel billets in about four to five pieces each. The average knife has about 280 to 400 layers, which you can get with about 18 to 22 sheets of steel.

Lots can also get Damascus with a nine-inch angle grinder. Besides, you can always cut and restack once the steel is hot enough. That is what you need on average, but you can try different combinations for your piece.

I always like to have a little bit more billets when making a bigger knife since it can thin out when you're hammering.

Plus, you can add more to future blades, especially when you want to use a particular steel for alloys.

Besides 1084, What Are Other Good Steels for Beginners?

If you are not doing the heat treatment at home, then the 440c is a good option since it will retain its hardness. Otherwise, O1, 1095, and 5160 are excellent alternatives to 1084. O1 is a nice choice, in particular, because you can get it at almost every machine supply place.

5160 is probably the best out of all of the optioned mentioned above because it is like working with real steel and creates usable blades. Furthermore, you do not have to make it super hot. You can even heat it in a kitchen oven if you are pressed for space.

What Stainless Steel Is Relatively Easy to Heat Treat?

CPM 154 is one of the best most accessible stainless steels to heat treat. It is not very expensive and has a superb quality. It also has a high temper temperature, so you can heat it without risking the quality of the blade.

AEB-L is another good stainless steel if you are working with a more basic setup. The reason is that it is incredibly forgiving and reasonably priced compared to other models. It works well with an alcohol quench and will give you an excellent basic blade, too.

When Using 01 Tool Steel, What RC Rating Do You Aim for After Quenching?

After quenching, you want to aim for the highest RC rating that you can get with this 01 tool steel. It typically means that you should go for about an RC of 63-65 for best results.

Some people get good results with a 61-63 if they use 01 tool steel with things that won't snap it. In a few cases, small cutters do well with a 59 rating even. Still, keep in mind that their retention will be affected the lower it goes.

Having a higher hardness level always means higher wear resistance but less toughness. Meanwhile, lower hardness will mean less resistance to wear, but you will have a much tougher blade.

Can You Weld Damascus Steel to Standard Stainless Steel so That You Can Have a Damascus Blade and Hide the Weld Under the Knife and Save on Costly Damascus Steel?

That depends on your materials. Damascus steel has an extremely high carbon composition. You need to combine it with something that complements its carbon content.

Sometimes, some people will pattern out to avoid spending much on Damascus steel. The problem is that it can create an unbalanced pattern of hardness to the blade when you create it.

You can use pattern welding a lot more these days because it was initially a solution to older blade problems in the past. However, know that this can be used if the blades have a good carbon composition together and there are other metal alloys to compliment this.

CHAPTER 5: RECYCLED STEEL

R ecycled steel has specific properties that other types of steel may not have.

In this chapter, you will learn about some of the different aspects of recycled steel. I will also try to answer some of your questions about recycled steel and why it matters.

What Should I Do With Leaf Spring That Is Too Wide to Use? Should I Hammer It to a Usable Width or Cut It Up?

Typically, leaf springs are similar to 5160 when used to make knives. It can be great steel, even if it is incredibly tough. When you test the file, it should be fine if it goes around like glass. If not, hammering will give you the size that you want.

I do not suggest grinding or cutting leaf spring. After heating and quenching it, it can be challenging to grind and cut. This type of steel is better to create by hammering instead of grinding to speed up the process.

Is Leaf Spring Good for Making Knives?

Leaf spring is one of the best types of recycled steel. They are extremely easy to get. You can also straighten it, cut what you need, and start making your knives. Hammering is a better option for leaf spring steel.

A little bit of leaf spring goes a long way. You do not need a ton of it to make a great knife. You should understand that it is a hardened steel, so it will have to be slowly cooled. However, it creates good knives, and 5160 steel is a good option if you need something that works with it effectively.

I have worked with leaf spring before and notices that it takes a lot more time to use than other types of metal. I usually let it quench for a little while longer too to keep its hardness level.

Is There a Special Way to Harden Circular Saw Blades?

In general, grinding circular saw blades with a ¼ inch wheel grinder gives you the ultimate control with the knife itself. As for heat treatment and tempering, you should heat the edge first to the nonmagnetic level. Then, you can quench it in vegetable oil.

When tempering, always take the correct precautions and wear leather all over. After that, clamp the blade to the vice.

Temper the blade to about 350 to 400 degrees for a few hours. The thing is, saw blades are not the best for this technique. Still, some of them come in 5260 versions, and you can get something good for heat treatment.

What Is a Spark Test? How Do I Do It?

Spark testing is how you figure out what type of recycled metal you have. You do it by taking a sheet and literally scraping it against the grinder.

You watch out for the sparks as they will tell you the ferrous metals that you are working with. It is simple, quick, and great for people who do not want to spend forever trying to know what metals are in their hands. After that, you need to consult a chart.

The one downside to this type of testing is that it can be hard to figure out the exact heat treatment and carbon in the metal. Still, it is useful if you want to get a general idea of what you are working with.

How Can I Fix Uneven Bevels?

Dealing with uneven bevels is a problem that can sometimes happen with recycled steel. Often, you are working with something that you may not know the full properties of. The only thing you can do is start with the higher angles and then go lower.

Try to reprofile a knife every so often, looking it over to see how it looks. When I work with an uneven bevel, I make sure to check it after every cut. Monitor the edges constantly when you are doing it since you might take away too much if you are not careful.

As a reminder, you should take things slow and do not try to do it all at once. Avoid wasting your time for pure symmetry. Instead, work from one side to the other before putting everything together. Furthermore, your grip and hand placement can affect the evenness of a bevel.

Some people get obsessed with perfecting bevels, which may be fine in some ways. But in other cases, leaving a knife a bit uneven may not be the worst decision. If you are working with kitchen knives especially, one side may be a little sharper than the other for cutting purposes.

Is Tool Steel (W1, W2) Good for Making Knives?

W1 is an excellent steel for making knives since it can be tempered and etched quite easily. However, W2 is a better type of steel than W1.

The reason is that W1 is water hardening and has a much higher level of carbon content. In contrast, the W2 is a hard steel, but its edges are not anywhere near as robust as how W1 can be at times.

W2 is a good one for longer knives and thicker spines. Some believe that it is almost as tough as 5160, but that's debatable. Personally, I like to use W2 instead of W1 on my knives, but I have made decent blades with both types.

Overall, tool steel is terrific for making knives and a good beginner steel.

I Have a Blade With an Acid Finish. How Do I Protect It From Rust?

Rusting commonly happens with blades from metals that you do not know about. The key is to keep other acids away from the edge. Wiping it down with oils can also do the trick.

Acid etching also forms a patina that can be used to help prevent rust from forming. The patina is a black layer of rust, which is much stronger and more resistant to corrosion than the other types that you will see on other blades.

If you keep the blade clean and wash and dry it after use, though, you should be fine. The upside is that cutting fruits and veggies will help the knife form the patina.

How Do I Figure Out the Rockwell Hardness of Steel Without Using Special Tools?

The Rockwell scale hardness test is the most accurate type out there. They are simple to do, too. The only downside is that it works best with a diamond or something hard.

Not all of us have a diamond lying around, but an excellent way to do this is to imprint a diamond into the center of steel and measure how deep this goes. This is how lots of people figure out the hardness of their materials.

Another way is by using a ball of steel. If you do not own one, you can take some steel, form it into a ball, and then press it into the center. However far it goes is, of course, how hard this type of steel is.

From there, you can determine whether or not the steel's hardness level. There is a scale for you to tell that, and you can consult a chart as well. That way, you know just how hard the material is and what you can do with it.

You can also use the RC hardness scale along with the spark test to determine the quality of your metal. Understand, though, that it does not fully replace chemical testing.

Make sure not to spend some time looking at the chemical composition of steel if you can. If you are in complete doubt on how to do this, you can contact the manufacturer and find out whether or not your steel has specific properties or not.

How Can I Keep Knives Symmetrical?

The most obvious way is to do the same thing on both sides of a knife. For example, if you sharpened it about five times on one side, you should do the same on the other side. Grind it to 50/50 as well and always check each side.

When you are cutting or creating scales, you should always work at equal angles. This way, you can get the same effect as you did with the other side of the knife.

Remember that the angle you are using, and the pressure exhibited on your knives can keep them perfectly balanced and symmetrical. If you do not do that, you will end up with crooked knives.

The process can be a bit frustrating at first, but the best way to do it is to try it repeatedly. Your muscle memory can develop from this so that you can make better knives each time.

Usually, working with your knives can help you fix their symmetry. But if you need a second pair of eyes to make sure that the blade is symmetrical, someone can do that for you.

After every symmetrical cut and grind, you should check the blade and see how exact it is. This takes time, but working with symmetry allows you to get the most out of your knives.

CHAPTER 6: MISCELLANEOUS

We tackled a lot of the crucial questions about forging steel, but there are still some left unanswered.

In this chapter, we will talk about other blacksmithing techniques that you need to understand.

How Are Traditional Katanas Different From Modern Katanas in Terms of Techniques?

New katanas are more homogenous and predictable than traditional ones. In the past, they forged katanas by the "way the moon looked during an evening." However, nowadays, there are more precise means to make them.

Modern blades are often thicker, a bit heavier, and more tapered at a different length. They are also more heat-treated, which means that they are more durable than the previous models.

The problem with modern blades is that some of them are made with cheap materials. There are also reports of people breaking them on water bottles.

While modern blade may look like they are stronger, that is not always the case when you put them to the test. I do like to examine any knife that I make to see if they can be used for decorative reasons.

How Does a Replica Katana Differ From the Modern Version?

To say that it is all in the steel is not 100% true, but it has a significant effect on the blade's quality. Most replicas are made with stainless steel, so they look pretty. However, they are not as strong as the traditional ones.

Traditional swords use carbon steel, but their quality suffers due to the impurities. So, they are nowhere near as strong as a modern sword in some regards.

Most swords used for function rather than decoration will have carbon steel. Others, such as the legendary Damascus sword, use a type of steel that is similar to the traditional Japanese katanas, which are also made out of carbon steel.

How Do I Get Selected for Forged in Fire?

Forged in Fire, if you do not already know, is an incredibly popular bladesmithing show. To get selected for this, you need at least two years of forging experience. You should also be able to create a blade that is at least 15 inches long.

You need to contact the team to get on the show. However, keep in mind that they want to find female smithers.

I think that it is a great show. If you would like to show off your skill, this is a beautiful opportunity to do so. However, they may take a while to get back to you. If you do not hear from them at once, try not to get discouraged.

What Kind of Epoxy Should I Use for a Knife Handle?

It depends on what you are going for. On average, kitchen knives go well with the G-flex epoxy, and most knifemakers benefit from using this.

West Systems is known for making good epoxy that holds the bonds nicely, too. If you are in the market for some epoxy for your knife handles that has a good curing time and will last a while, this is your best bet.

You will probably need to look online to find a distributor, though.

I Cannot Find the Head Diameters for Screws that I Bought for My Folding Knives. What Should I Do?

The best option is to contact the supplier directly. Emailing the company will help you figure out the diameter. If you cannot reach anyone, though, you should take your calipers and measure it.

I like to find out the diameter by doing the latter and then order the right screws if I have the wrong size. Usually, doing it yourself takes far less time, but I know some people do not have the skill to do so. Thus, a quick email usually suffices.

What Type of Equipment Setup Should I Use to Start Heat-Treating My Knives?

Some types of steel can be treated in an oven, so you can heat it to 400 degrees. The temperature is anywhere from 350 to 450 degrees, which is not that hot but just enough to soften it slightly.

You will still need a fireproof quenching container. Do not use Tupperware for this because that's a recipe for disaster. You can heat the oil to help with quenching using a hot plate. Others use a kitchen toaster for the tempering step at the end.

I like to use a ceramic basin for heat treatment since it can withstand incredibly high temperatures. But then again, that is up to you. It is also easier to get ahold of, and ceramic tends to be stronger than other types of medium.

Sometimes, if you want to heat your knives further, you can get a blowtorch. Basic steel reacts in some way to this type of heat, so it is worth checking out.

The Shop Dust Has Finally Got to Me. Do You Have Any Recommendations for Buying Quality Air Filter or Dust Collection System?

Dust collection is significant for keeping your shop safe to use. Too much dust can get into your lungs, causing respiratory issues with all the chemicals in there.

The JET DC-1100VX-CK Dust Collector is a good option if you want a full-on dust collection system that runs at 1.5 amps. It does not require much power and has its cooling system.

This is great if you are working with more large-scale projects and need something to clean up the area quickly. However, it does cost a bit more than most want to pay.

Air filters can be cheaper than a dust collection system, meanwhile. The Honeywell 50250-S True Portable Air Purifier is an excellent choice for that. You can plug it into an outlet in your shop or garage, and the air particles will be sucked up and filtered out.

I do recommend getting either or both if you plan on black-smithing seriously.

How Do I Etch My Marks Into My Knife?

Electro-etching is the most popular way to etch your marks on a knife. However, it can cost almost $200 online. You can make your own with a DC transformer, a lamp or speaker cord, alligator slips, stainless steel, etching solution, and masking vinyl.

To make this your tool, you can hook the wires to the transformer and wrap the plate into the felt. The latter works as the sponge for the etching solution. Prepare the blade by masking that areas that won't get etched.

Add the ground clamp to the blade, wet the felt, and then rub the edge with the positive stainless steel plate. It should etch in about five to 10 minutes. Wipe away the excess sludge and make sure that the masking vinyl is not lifted when working. Otherwise, the edge won't be clean.

Do Most People Think That the Outer Faces of a Sawn Pair of Natural Scales to the Desired Handle Width and Mount the Adjacent Inner Sawn Faces to the Outside to Mirror Each Other?

Some people do that. Most of the time, if you want an exact and detailed length to these that is accurate, I recommend it as well. However, others do it with a thicker outer face.

Remember that doing so can blunt the knife a little bit, and it is not the best strategy. Usually, thinning the handle creates a more exact, identical look to the blades.

How Many Knives Should I Display When Trying to Sell Them at a Convention?

It should be 31 knives. The number might be a little bit random, but it allows people to see your selection enough and judge the craft. Plus, you can make multiple variations to attract the attention of convention-goers.

I sometimes like to take 30 to 40 knives to conventions and make sure that they are of different sizes. You should never take all the same blades. Do your research on which knives are best for every convention as well.

What Tools Do I Need to Polish My Scales?

The same methods used to polish the steel can be used to clean your scales.

You can take a shortcut and polish it with 800-grit sandpaper first. Then, spray it with a glossy polyurethane over the scales to give a beautiful look. It is faster and may be suitable for your scales if you are pressed for time.

I like to use the 800 myself. But if you use the 1000-grit paper to polish the scales, it can make your blade look very smooth.

I also use the 400 grit and then the 200 grit to thoroughly polish it off. Doing so can make your blades smooth with the right care. Sometimes, the extra 200 will give it a shine that helps improve the look to the scales too.

When Filling the Bevel, Should I Use Oil?

Some people may think that filling the bevel will help you get a better look and make it more even. However, draw filing is a much better option and allows for a more exact look that you are going for.

Oil may help, but it creates a lack of friction in between, so it may not be useful. I personally polish off the scales sometimes, but I also like to keep their edges rough.

Hunting knives need a bit more polishing than others. If you are using it for cutting, you need to keep the blades nice and smooth.

For decorative blade, you can get away with a rougher texture. If you do not want to sand or oil them up as much, you do not have to.

I Struggle With Dyeing My Leather Sheath. Are There Any Tips on How to Dye It Evenly?

First, make sure that the leather you are using can take the dye. Some types of leather cannot get dyed, after all.

Next, apply your dye in circular motions. That way, it will cover all areas of the sheath evenly. Apply a coat diagonally to the right and then do the same to the left to even out the dye and make it look good.

Finally, you want to apply your finish and buff it lightly. From there, you should have a perfectly dyed leather sheath. You can add more dye to make it darker, too.

I Want to Make a Sheath Out of Kydex. What Is the Standard Size for a Kydex Sheet?

On average, 0.93 is suitable for the larger blade sheaths. However, if you are working with a smaller item, you want to use the 0.80 version. For the latter, always know that you will need the ¼" rivets for this type of sheath.

If you do not like waiting, you can even go up to 0.60 since it heats up pretty fast. Just know that you will need to use ⅜" rivets for good results, though. 0.93 is good, but it can be too thick, even for the larger blades.

Usually, anything over 0.60 can be a bit much for most blades. Still, it all depends on what type of blade you are trying to go for at the end.

Is It Normal to Apprentice or Mentor Under Another Forger?

Yes, that is normal for people to do. After all, more experienced individuals can help teach you tricks and shortcuts that improve your forging skills.

An apprenticeship can be helpful for many who want to learn blacksmithing techniques. Personally, I was an apprentice for a while until I took forging into my own hands and learned the ropes.

With mentors, you can learn valuable skills immensely.

However, I also know that some people do not work well with mentors. It can ultimately bring forth bad habits, so you have to be careful.

If you work with a mentor or enter an apprenticeship, you should do it with someone who knows what they are doing.

How Do I Avoid the Epoxy Mess When Gluing Scales?

Epoxy is very messy and challenging to get off. To start, I suggest working slowly and making sure that you clean any excess off immediately before it hardens.

If you end up having excess epoxy on the item after it hardens, you can take some sandpaper and scrape it off. You need to take your time with it and understand that going too hard will result in scratches on your knife.

You can also sand and buff after you do the epoxy to eliminate a lot of the mess. However, not everyone likes to do that since it can be hard to make it look good before you put the epoxy on there.

Some people use brass rods or dowel to scrape up the epoxy after it hardens. The best rule of thumb is to make sure that you are taking it slow. If a mess happens, clean it up right away.

Should I Stabilize the Wood Before Using It as a Handle?

Yes, you should. Unstabilized wood comes apart if you use it as a handle if you are not careful.

Many people will marvel at the knives that they have made, only to cry out in frustration when they realize that they have forgotten to stabilize the wood. Now, their knife flies out, and the handle comes apart.

Always make sure that the wood is stabilized and put together before you use it as a handle. Doing otherwise can make your blade awkward, and it may unravel when you are using the knife.

What Sort of Grit Do I Need for a Mirror Polish?

Some people get an excellent mirror polish with an 8000 grit and then go to 4000 and then 2000 before buffing it. However, you will need to do the 8000 grit once again for the rest of the knife's life. It creates a more satin than a mirror polish.

Meanwhile, a 1500 or 2000 grit can give you an incredible mirror shine. You will need to hit that with some polishing paper once it is done.

It also works well if you use a dykem over the beginning and then work with an 800 grit by machine and move it up to 1000 or 1500 grit by hand. With this method, you do it at 1000 at 45-degree angles and then proceed to 1500.

What Is a Good Price Point for My Knives?

The simplest way to figure out the pricing is to know how many hours you have put into it, as well as the price of your knife. From there, double the hours that you have worked.

Let us say that you worked for about 5 hours on a knife, and you want to be paid 12 dollars per hour. That will be $60 just for the labor, but you can double that to $120. In case the materials cost you $20, you may ask for about $140 for the knife.

However, that price may scare off a few people, so you need to be realistic on how much you want to be paid for your labor.

Sometimes, if it does not sell, you may drop it. You are still turning a profit, albeit not as much as you think you would.

If you are trying to get your name and brand out there, I suggest being more modest with the pricing. After all, your knives are not super popular yet. You need to understand that when it comes to generating a profit, sometimes you will need to lowball it.

But of course, never sell yourself short. If you want to cover the materials and the labor minimally, do that. Otherwise, the double labor plus materials cost is reasonable if you have generated a bit of popularity with your materials.

Also, look at what the pros are selling the knives for. If someone is settling a small knife for $100, and they have done this for years, selling yours for about $50 or $60 may be right. You should look at competitors, see what they are doing, and price yourself relatively from there.

How Can I Preserve Wooden Kitchen Blade Handles?

Using Ballistol oil is one of the best ways to preserve your knife handles. Rubbing it on a few days after making your blade will help to maintain it. Besides, it creates a beautiful, even surface that does not break down.

Ballistol oil, in particular, is incredibly sturdy. It can be used to keep your piece looking nice and shiny. You want to continue to apply the oil in small drops and add it until you notice that the wood is not soaking up any more of the oil.

Danish and Linseed oil can also preserve kitchen blade handle well since they have a pleasant smell and excellent properties.

Some people think that olive oil is a great idea, too. While it does take care of the piece well, the problem is that its smell becomes rancid. It can be a bit pungent after a while.

At What Point Do You Stop Calling Them Tangs and Scales and Start Calling Them Knife Handles?

This is a good question. Often, people will call knife handles as tangs and scales until the assembly happens. For example, when the screws are in place, and everything is put together, then you start calling it as a knife handle.

However, most individuals still refer to the tang as a part even after it has all been assembled because that is a part of the knife. Handles are more of a colloquialism than the tangs, and the latter are a big reason why people choose a knife.

So, if you are around knife experts and want to sell your blades to them, you should become familiar with the terminology. Scales are usually associated with the tangs when it is all infused together but expect to hear the term more when you start assembling the knife.

What's the Best Pocket Knife Brand?

The best pocket knife brands in my opinion are Buck, Kershaw, Spyderco, and Benchmade Mini Grip. The latter comes under 60 bucks, to be specific, and it can lock the blade as well. However, you may need to sharpen the knife after receiving it.

All of these brands are excellent because they can provide the exact types of knives that you may need. Plus, for the price point, they definitely are good if you want something effective and affordable.

How Do I Get a Good Shipping Rate on Knives I Sell?

First and foremost, never ship your knives through USPS. Even when you insure your packages, it can get lost, or worse, tampered with.

I suggest shipping knives in recyclable bubble fillers, wrapping them in the packages, and putting one of those edge guards to keep them from ripping accidentally.

You will want to get insurance when you ship knives. If you do not, you may end up losing a lot of money later on, which may affect your business. If you export in bulk, though, you may be able to get a reasonable shipping rate.

In general, being honest with what is in the package will help you more than anything. After all, people will be much more interested in what you are shipping, so they can keep it safe.

Putting extra packaging and filler is good for the knife, too. If you are lucky, you can ship it in those fillers. Just make sure that they are adequately packaged so that it can survive the apocalypse.

These days, it is hard to get reasonable shipping rates, but figuring out the shipping costs and adding that to your price, along with insurance, will help.

Should I Stabilize Hardwood to Make Knife Scales?

That depends on the type of hardwood you are using. Naturally, oily ones do not need that extra stabilization. Some of them won't even take the excess oil, and it can split the wood violently, which is what you do not want.

The ones that you should not add stabilizer to include ironwood, cocobolo, African blackwood, ebony, rosewood, Bubinga, and walnut.

If you are using another type of hardwood, the best way to know if you can use a stabilizer is to try and see if that takes the oil.

What Is the Best Way to Make a "Chipped Blade" Style?

Your best bet for that is a Dremel. It lets you get that chipped look quite quickly. Dremels are a handy tool when you are smithing and make an excellent hardware tool.

Despite that, I will say that going over the blade with a Dremel a few times every now and then might be good since it can get nicks and scratches. Chipped blades also need to have a rougher edge to them, so keep it in mind when you put the knife together.

How Do I Put a Logo on My Blades?

Vinyl cutters are your best friend for making stencils for etching. They are easy to procure, cheaper than laser alternatives, and can be left right there. Vinyl stickers with ferric are good as well, and the stencils can hold up on their own.

Along with that, the 12-volt etching system can help you etch logos onto blades. Consider this option when you are trying to put logos on whatever you are making.

Should I Offer a Warranty on the Knives I Sell?

I highly recommend offering a warranty so that you can send out a replacement if the customer breaks it. The thing is, most people will try to take care of their blades instead of ruining it. So, you may not even need to replace any knives.

Understand that you can never produce a foolproof knife. Someone will always figure out a way to do it. However, if you buff it up, sharpen the edges, or even make sure that the scales are not removable before sending it out, then you can reduce the need for warranties.

Plus, you can choose for yourself how long you want the warranty to be. Having a warranty in place for the rest of your life does not sound like the best idea because people may come to you far down the road after you have stopped setting up shop and demand for a replacement.

I always make sure to include a warranty on my knives. So far, I have only had two shipped back to me because they were my first knives. I eventually managed to fix them quickly.

If you include instructions on how to take care of the blade, people will follow them. They know that it is better to take care of it now than have to send it back.

What Separates a $100 Knife From a $3000 Knife?

Usually, brand names make a knife expensive. The manufacturer may also use certain materials and finishes on them. Some of them may even end up emulating older styles that you may not have seen before.

A few expensive knives also have custom finishes and fittings for the customer at hand. The polish and the pride of ownership play a role in this, too.

In terms of effectiveness, it all depends on how the knives have been made. Some $3000 knives have a very high tolerance compared to what you may see from other makers. If you do not care much about the brand, though, then a $100 should suffice.

If you want to take pride in making something pricey, you can do it. Other people may even pay thousands of dollars for the name alone. If you have the skill, and your name is known, you can charge more for this.

Some makers will charge an arm and a leg for their knives only due to their brand. If you can justify a higher price, then go for it by all means. Otherwise, slowly work your way up, and do not be afraid to charge a bit more for customized knives.

What Is a Good Combination for Making the Paste Used for the Hamon on a Katana?

An excellent combination to make the paste is white vinegar and 1500 silicon carbide abrasive.

To begin, sand the blade with a 1500-grit paper and soak it in white vinegar. You may etch the object as needed and then use the 1500 silicon carbide on it.

Proceed to go back and forth between these two materials. You can then polish the blade further with lemon juice and neutralize the acids with Windex later.

Another popular solution is the 8:1 ratio of water and ferric chloride. After doing it a few times, you can neutralize it with some simple green before washing it off. This can polish and make a nice Hamon on a katana.

If you want a smokier or grayer look to your Hamon, then I suggest the first combination since it provides a cloudier appearance. The best type of steel to use to create a good Hamon is W1 steel.

These are usually the top questions that people have when dealing with knives and blades.

CONCLUSION

And there you have it, 101 basic questions about bladesmithing that everyone who starts out asks.

I understand that it can be complicated to begin with. However, with this information, you will be able to hone your blade skills and make yourself the best smither possible.

Many people who start with bladesmithing struggle with these basic problems with their blades. By understanding what you are doing with the blade and crafting them however you want, you will be able to create the best knives.

I know it takes a lot of skill and practice to create knives, and a lot of beginners are intimidated by the sheer process alone, along with the amount of materials they need. But as you can see, it is easy to get what you need, and it does not have to cost you a lot of money.

I hope that this bladesmithing book will help you understand what you have to do in order to ensure successful forges. If you ever want to learn bladesmithing techniques, now is the best time to do so.

Go forth and start making some great knives that you can use today!

REFERENCES

Aalders, H. (2017). Spark Testing Mystery Steel for Knife-making (with Walter Sorrells video). Retrieved from http://www.thetruthaboutknives.com/spark-testing-mystery-steel-for-knifemaking-with-water-sorrells-video/

Alpha Knife Supply. (n.d.). W1/W2 Information and composition. retrieved from https://www.alphaknifesupply.com/zdata-bladesteelC-W1-W2.htm

Comeau, D. (n.d.). DIY Micarta. Retrieved from http://dcknives.blogspot.com/p/blog-page_6.html

Fischer, J. (n.d.) Knife Anatomy, Parts Names By Jay Fisher. Retrieved from https://www.jayfisher.com/Knife_Anatomy_Parts_Names_Definitions.htm

Haddad, K. (n.d.). Beginner Knife Making Equipment Guide. Retrieved from https://www.tharwavalleyforge.com/articles/tutorials/102-beginning-knifemaking-equipment-guide

Knifepath. (n.d.). What's the Best Oil for Quenching Steel? Retrieved from https://www.knifepath.com/best-oil-quenching-steel/

Knife Up. (n.d.). CPM 154 Knife Steel overview. retrieved from https://knifeup.com/cpm-154-knife-steel-overview/

Knife Up. (n.d.). N690 Steel Properties. Retrieved from https://knifeup.com/n690-steel-properties/

MrBalleng. (n.d.). Adding Serrations to a Pocket Knife. Retrieved from https://www.instructables.com/id/Add-Serrations-to-a-Pocket-Knife/

Newage Testing Instruments. (n.d.). Rockwell Hardness Testing. Retrieved from https://www.hardnesstesters.com/test-types/rockwell-hardness-testing

Nowlan, P. (2017). Sharpening School Lesson #2: Grasping the Sharpening Fundamentals. Retrieved from https://www.knifeplanet.net/lesson-2-sharpening-fundamentals/

Sharpening Supplies. (n.d.). Detailed Discussion on knife Sharpening Angles. Retrieved from https://www.sharpeningsupplies.com/Detailed-Discussion-on-Knife-Sharpening-Angles-W28.aspx

Smith, J. (2018). How to Temper Knife Blades. Retrieved from https://careertrend.com/how-4897436-temper-knife-blades.html